TIRE PRESSURE SPECIAL STUDY

DATA DOCUMENTATION

U.S.. Department of Transportation
National Highway Traffic Safety Administration
National Center For Statistics and Analysis
Washington, D.C. 20590

Introduction

Background

In 2000, Congress passed the Transportation Recall Enhancement, Accountability, and Documentation (TREAD) Act. Section 12 of this act directed the Department of Transportation to complete a rulemaking requiring implementation of a warning system in new motor vehicles indicating under-inflated tires.

In response to Section 12 of the TREAD Act, NHTSA's National Center for Statistics and Analysis (NCSA) conducted the Tire Pressure Special Study (TPSS). The TPSS was designed to assess to what extent passenger vehicle operators are aware of the recommended tire pressures for their vehicles, the frequency and the means they use to measure their tire pressure, and how significantly actual measured tire pressure differs from the manufacturer's recommended tire pressure.

Data Collection Methodology

Field data collection was conducted through the infrastructure of the National Automotive Sampling System Crashworthiness Data System (NASS CDS). The NASS CDS consists of teams of researchers located at Primary Sampling Units (PSUs) throughout the United States. The PSUs are located in urban, suburban, and rural settings with nationally representative populations.

The population surveyed by the researchers in the TPSS represents a sample frame consisting of drivers who used gas stations to refuel their vehicles between the hours of 8:00am and 5:00pm. Data collection was conducted from February 1, 2001 through February 14, 2001.
Vehicles surveyed included passenger cars and light trucks. NHTSA classifies light trucks as utility vehicles, light conventional trucks and van based light trucks with a Gross Vehicle Weight Rating less than or equal to 10,000 pounds. A total of 11,530 vehicles were included in the survey, of which 6,393 were passenger cars, 1,894 were utility vehicles, 1,385 were van based light trucks, and 1,857 were light conventional trucks. The distribution of vehicles is consistent with vehicle registration.

Data collected during the TPSS included daily site information, driver interview and profile data, vehicle profile data, and tire data for all four tires on the vehicle. The vehicle information collected included vehicle profile data and the manufacturer's recommended tire pressures. Tire information collected included tire profile data as well as air pressure, sidewall temperature and tread depth measurements. A complete description of the data collection process is presented in the NHTSA Research Note "Tire Pressure Special Study: Methodology."

Sampling and Weighting

Sampling for TPSS was performed in four stages. The first stage of sampling was the NASS CDS Primary Sampling Units (PSUs); the second stage was selection of seven Zip codes within the PSUs; the third stage was the selection of a random sample of two gas stations within each Zip code. The final stage of sampling was the identification of four motor vehicle body types: passenger cars, utility vehicles, van based light trucks, and light conventional trucks. Weights were produced for all sample vehicles accounting for each stage of the sample design. The TPSS data file includes the SAS variable WGT, which will produce weighted national estimates.

Introduction

File Structure

TPSS data is presented in two SAS files: the Vehicle file and the Tire file. The Vehicle file contains one record per vehicle and has 11,530 records. The Tire file contains one record per tire and has 46,120 records. The two files can be linked by the variable CASENO, which appears on both files. Both the Vehicle and Tire files contain variables describing tire size. The tire size variables on the Vehicle file refer to manufacturer recommended size found on the vehicle placard or owner's manual and have the prefix "V12_"; the tire size variables on the Tire file refer to the size of the tires observed on the vehicle and have the prefix "T_".

Missing Data

Missing numeric data is represented by the SAS special code of ".". Where character data is missing the fields are left blank.

Tire Size Formats

There are at least three different tire size formats used by tire manufacturers. Each format is composed of multiple components of tire size, and not all formats include the same components. Most of the tire sizes in the TPSS are in accordance with the three major formats. All other formats were grouped into an "other format" category. Examples of the three formats are shown below. The components, and possible values for each component, are discussed in the following sections.

Format	*Example*	*Components*	
Metric	P205/75HR14	P	= Passenger Car Tire
		205	= Section Width in millimeters
		75	= Aspect Ratio
		H	= Speed Rating
		R	= Construction Type
		14	= Rim Diameter in inches
Light Truck High Flotation	31X10.50R15	31	= Tire Overall Diameter in inches
		10.50	= Section Width, in inches
		R	= Construction Type
		15	= Rim Diameter in inches
Light Truck Numeric	8.75R16.5	8.75	= Section width in inches
		R	= Construction Type
		16.5	= Rim Diameter in inches

In the TPSS data file the various size components are coded separately. That is, for metric format tires, each of the six components shown in the example above is in separate variables. For light truck high flotation tires the size components are in four variables, and for light truck numeric tires the components are in three variables. In some cases one or more of the components are missing due to difficulty in reading the sidewall lettering.

Variable List

Variable List: Vehicle / Driver Level Variables:

SAS Name	SAS Label
CASENO	Case number
PSU	field collection PSU
PSU_STR	stratification level for PSU
SITE	field collection Site
DAY	field collection day, February 2001
OBS	field collection observation number
T05	ambient air temperature, inspection site
D05	maintaining proper infl. a concern
D06	miles traveled
D07	responsible for maintenance
D08	primary driver of vehicle
D09	driver response, rec. tire pressure
D10	method of determining proper pressure
D10S	question D10 – other, specify
D11	method of checking actual pressure
D11S	question D11 – other, specify
D12	frequency of checking pressure
D12S	question D12 – other, specify
D13	gender
D14	race
D15	age group
V05	vehicle model year
V06	vehicle make
V07	vehicle model
V07S	vehicle sub-model
V08	vehicle body type
V09	vehicle identification number
V10	GAWR, front
V11	GAWR, rear
V12_TYPE	manuf. rec. tire type
V12_C	manuf. rec. tire construction
V12_SR	manuf. rec. tire speed rating
V12_MT	manuf. rec. metric tire type
V12_MW	manuf. rec. metric tire width
V12_MA	manuf. rec. metric tire aspect ratio
V12_MR	manuf. rec. metric rim diameter
V12_HD	manuf. rec. flotation tire height
V12_HW	manuf. rec. flotation tire width
V12_HR	manuf. rec. flotation tire rim diameter
V12_NW	manuf. rec. numeric tire width
V12_NR	manuf. rec. numeric tire rim diameter
V12_OTH	manuf. rec. other tire size
V13	manuf. rec. pressure, front, cold
V14	manuf. rec. pressure, rear, cold
V15	manuf. rec. pressure, front, hot
V16	manuf. rec. pressure, rear, hot
WGT	case weight

Variable List

Variable List: Tire Level Variables:

Name	SAS Label
CASENO	Case number
POSITION	observed tire position on vehicle
T_MANF	observed tire manufacturer
T_TYPE	observed tire size format
T_C	observed tire construction
T_SR	observed tire speed rating
T_MT	observed metric tire type
T_MW	observed metric tire width
T_MA	observed metric tire aspect ratio
T_MR	observed metric tire rim diameter
T_HD	observed flotation tire diameter
T_HW	observed flotation tire width
T_HR	observed flotation tire rim diameter
T_NW	observed numeric tire width
T_NR	observed numeric tire rim diameter
T_OTH	observed other tire size
MAXP	observed tire maximum pressure
MSP	observed tire measured pressure
TEMP	observed tire temperature
TREAD	observed tire tread depth
WGT	case weight

Vehicle Level Variable Attributes

SAS Name: CASENO   **SAS Label: Case number**

Type: Character

Length: 8

Remarks: This variable is a unique identifier for each vehicle in the data file. It is an 8-character field. The first two characters represent the PSU for the case, the second pair of characters represents the site in the PSU, the third pair represents the day the interview was conducted, and the last pair represents the individual vehicle for the PSU/Site/Day combination. The variable is on both the Vehicle and Tire level data sets and is used to link the two sets.

SAS Name: PSU   **SAS Label: field collection PSU**

Type: Numeric

Remarks: Primary Sampling Unit Number

Attributes: xx   PSU number

SAS Name: PSU_STR   **SAS Label: stratification level for PSU**

Type: Numeric

Remarks: PSU's are divided into 12 strata for calculation of standard errors.

Attributes: xx   PSU number

SAS Name: SITE   **SAS Label: field collection site**

Type: Numeric

Remarks: Site number within the PSU, based on Zip code selection

Attributes: xx   Site number

SAS Name: DAY   **SAS Label: field collection day**

Type: Numeric

Remarks: Date in February when fieldwork was conducted

Attributes: xx   Day

Vehicle Level Variable Attributes

SAS Name: OBS   **SAS Label:** field collection observation number

Type: Numeric

Remarks: Observation number of the vehicle for the particular PSU/Site/Day

Attributes: xx   Observation number

SAS Name: T05   **SAS Label:** ambient air temperature, inspection site

Type: Numeric

Remarks: Ambient air temperature at time of tire inspection. Measured and recorded by the interviewer for each vehicle inspected.

Attributes: xxx   degrees Fahrenheit

SAS Name: D05   **SAS Label:** maintaining proper infl. a concern

Type: Numeric

Remarks: This variable contains responses to the driver interview question "Is maintaining proper tire inflation a concern for you?"

Attributes:
- 1   No
- 2   Yes

SAS Name: D06   **SAS Label:** miles traveled

Type: Numeric

Remarks: This variable contains responses to the driver interview question "How many miles did you drive to reach this destination?"

Attributes
- 1   1 - 3 miles
- 2   4 – 10 miles
- 3   11 – 20 miles
- 4   Over 20 miles
- 5   Unknown

Vehicle Level Variable Attributes

SAS Name: D07   **SAS Label:** responsible for maintenance

Type: Numeric

Remarks: This variable contains responses to the driver interview question "Are you responsible for the maintenance of this vehicle?"

Attributes: 1   No
2   Yes

SAS Name: D08   **SAS Label:** primary driver of vehicle

Type: Numeric

Remarks: This variable contains responses to the driver interview question "Are you this vehicle's primary driver?"

Attributes: 1   No
2   Yes

SAS Name: D09   **SAS Label:** driver response, rec. tire pressure

Type: Numeric

Remarks: This variable contains responses to the driver interview question "What is the vehicle manufacturer's recommended tire pressure for your vehicle?" If the driver checked the owner's manual or other source to answer this question the interviewer was instructed to code "Does not know." This question was skipped if the answers to questions D07 and D08 were "no".

Attributes: 1   Does not normally drive this vehicle
2   Does not know
xx   Tire pressure in psi

SAS Name: D10   **SAS Label:** method of determining proper pressure

Type: Numeric

Remarks: This variable contains responses to the driver interview question "How do you **normally** determine what pressure to set your tires?" This question was skipped if the answers to questions D07 and D08 were "no".

Attributes: 1   Owner's Manual
2   Vehicle Placard
3   Tire Labeling
4   Visually
5   Other (specify)
6   Does Not Know
7   Other Person Maintains
8   Unknown

Vehicle Level Variable Attributes

SAS Name: **D10S** **SAS Label: question D10 – other, specify**

Type: Character

Length: 25

Remarks: If the driver responded "other" for D10 (How do you normally determine what pressure to set your tires?) the other method specified by the driver is recorded in this field.

Vehicle Level Variable Attributes

SAS Name: D11 **SAS Label: method of checking actual pressure**

Type: Numeric

Remarks: This variable contains responses to the driver interview question "How do you **normally** check your tires for proper inflation?" This question was skipped if the answers to questions D07 and D08 were "no".

Attributes:
1. Visually
2. Pressure Gauge
3. Relative/Friend/Other person normally checks
4. Waits for vehicle servicing
5. Does not check
6. Other (specify)

SAS Name: D11S **SAS Label: question D11 – other, specify**

Type: Character

Length: 25

Remarks: If the driver responded "other" for D11 (How do you **normally** check your tires for proper inflation?) the other method specified by the driver is recorded in this field.

SAS Name: D12 **SAS Label: frequency of checking pressure**

Type: Numeric

Remarks: This variable contains responses to the driver interview question "How often do you **normally** check your tires for proper inflation?" This question was skipped if the answers to questions D07 and D08 were "no".

Attributes:
1. Weekly
2. Monthly
3. Whenever they seem low
4. When the car is serviced
5. When preparing for a long trip
6. Other (specify)
7. Does not normally check

NASS Tire Pressure Special Study

Vehicle Level Variable Attributes

SAS Name: D12S   **SAS Label: question D12 – other, specify**

Type: Character

Length: 25

Remarks: If the driver responded "other" for D12 (How often do you **normally** check your tires for proper inflation?) the other response is recorded in this field.

SAS Name: D13   **SAS Label: gender**

Type: Numeric

Remarks: Gender: Observed and recorded by the interviewer.

Attributes:
1. Male
2. Female

SAS Name: D14   **SAS Label: race**

Type: Numeric

Remarks: Race: Observed and recorded by the interviewer.

Attributes:
1. American Indian or Alaskan Native
2. Asian
3. Black or African American
4. Hispanic or Latino
5. Native Hawaiian or Other Pacific Islander
6. White

SAS Name: D15   **SAS Label: age**

Type: Numeric

Remarks: Age group: Observed and recorded by the interviewer.

Attributes:
1. Young Adult (16 – 24 years old)
2. Adult (25 – 69 years old)
3. Senior (over 69 years old)

Vehicle Level Variable Attributes

SAS Name: V05 **SAS Label:** vehicle model year

Type: Numeric

Remarks: Vehicle Model Year

Attributes: xxxx Actual vehicle model year

SAS Name: V06 **SAS Label:** vehicle make

Type: Character

Length: 25

Remarks: Vehicle Make: The vehicle make classifications are in accordance with vehicle make classifications used in other NCSA data including the General Estimates System and the Crashworthiness Data System.

SAS Name: V07 **SAS Label:** vehicle model

Type: Character

Length: 50

Remarks: Vehicle Model: The vehicle models are in accordance with vehicle model classifications used in other NCSA data including the General Estimates System and the Crashworthiness Data System.

SAS Name: V07S **SAS Label:** vehicle sub-model

Type: Character

Length: 50

Remarks: Vehicle Sub-Model: This information was not collected in the field and exists for only approximately one-third of the vehicles. The sub-model was occasionally obtained in the course of checking and editing the Make/Model/Vehicle Identification Number information recorded by the interviewers.

Vehicle Level Variable Attributes

SAS Name: **V08** **SAS Label: vehicle body type**

Type: Numeric

Remarks: Vehicle Body Type Category: Observed and recorded by the interviewer.

Attributes:
- 1 Automobile
- 2 Utility Vehicle
- 3 Van Based Light Truck
- 4 Light Conventional Truck

SAS Name: **V09** **SAS Label: vehicle identification number**

Type: Character

Length: 11

Remarks: Vehicle Identification Number: Observed and recorded by interviewer.

The vehicle identification number is a number assigned by the vehicle manufacturer. The VIN contains information on the vehicle such as manufacturer, model year, model, body type, restraint type, etc. For VINs with a length of more than 11 characters, any positions past the 11th character were deleted. The positions that were deleted contain the serial number, which can uniquely identify the vehicle.

SAS Name: **V10** **SAS Label: GAWR, front**

Type: Numeric

Remarks: Gross Axle Weight Rating (GAWR), front axle: Observed on the vehicle placard or the owner's manual and recorded by the interviewer

Attributes: xxxx   GAWR in pounds, front axle

SAS Name: **V11** **SAS Label: GAWR, rear**

Type: Numeric

Remarks: Gross Axle Weight Rating (GAWR), rear axle: Observed on the vehicle placard or owner's manual and recorded by the interviewer

Attributes: xxxx   GAWR in pounds, rear axle

NASS Tire Pressure Special Study

Vehicle Level Variable Attributes

SAS Name: **V12_TYPE**   **SAS Label: manuf. rec. tire type**

Type: Numeric

Remarks: Tire size format, vehicle manufacturer's recommendation. There are three commonly found tire size classifications for the types of vehicles in this study, each with its own format for presenting various size components. Tire size types that could not be easily categorized were placed in the "other" category. If more than one type of format was recommended by the manufacturer, the format that matched the format of the tires on the vehicle was recorded. See discussion of tire sizes in the Introduction for more information.

Attributes:
- 1   Metric format (example "P205/75/R/14")
- 2   Light Truck High Flotation format (example "31X10.50 R15")
- 3   Light Truck Numeric format (example "8.75/R/16.5")
- 4   Other

SAS Name: **V12_C**   **SAS Label: manuf. rec. tire construction**

Type: Character

Length: 1

Remarks: Construction type, all tire formats, vehicle manufacturer's recommendation. Observed on the vehicle placard or owner's manual and recorded by the interviewer.

Attributes: R   Radial ply tire

SAS Name: **V12_SR**   **SAS Label: manuf. rec. speed rating**

Type: Character

Length: 1

Remarks: Metric tire format: speed rating component, vehicle manufacturer's recommendation. The rating indicates the top speed for which the tire is certified. Observed on the vehicle placard or owner's manual and recorded by the interviewer.

Attributes:
- H   130 mph
- S   112 mph
- T   118 mph
- V   150 mph
- Z   over 150 mph

NASS Tire Pressure Special Study

Vehicle Level Variable Attributes

SAS Name: **V12_MT**   **SAS Label: manuf. rec. metric tire type**

Type: Character

Length: 2

Remarks: Metric tire format: vehicle type component, vehicle manufacturer's recommendation. This indicator is omitted from many metric tire formats. Observed on the vehicle placard or owner's manual and recorded by the interviewer.

Attributes: P   Passenger Car Tire
                LT   Light Truck Tire

SAS Name: **V12_MW**   **SAS Label: manuf. rec. metric tire width**

Type: Numeric

Remarks: Metric tire format: section width component, in millimeters, vehicle manufacturer's recommendation. The outer width of an inflated new tire from sidewall to sidewall. Observed on the vehicle placard or owner's manual and recorded by the interviewer.

Attributes: xxx   section width in millimeters

SAS Name: **V12_MA**   **SAS Label: manuf. rec. metric tire aspect ratio**

Type: Numeric

Remarks: Metric tire format: aspect ratio component, vehicle manufacturer's recommendation. The ratio between tire height and width. For example, an aspect ratio of 75 indicates a tire section height 75% as great as the width. Observed on the vehicle placard or owner's manual and recorded by the interviewer.

Attributes: xx   aspect ratio indicator

SAS Name: **V12_MR**   **SAS Label: manuf. rec. metric rim diameter**

Type: Numeric

Remarks: Metric tire format: rim diameter in inches, vehicle manufacturer's recommendation. Observed on the vehicle placard or owner's manual and recorded by the interviewer.

Attributes: xx   diameter in inches

Vehicle Level Variable Attributes

SAS Name: V12_HD   **SAS Label:** manuf. rec. flotation tire height

Type: Numeric

Remarks: High Flotation tire format: tire overall diameter in inches, vehicle manufacturer's recommendation. Observed on the vehicle placard or owner's manual and recorded by the interviewer.

Attributes: xx   diameter in inches

SAS Name: V12_HW   **SAS Label:** manuf. rec. flotation tire width

Type: Numeric

Remarks: High Flotation tire format: section width component, in inches, vehicle manufacturer's recommendation. The outer width of an inflated new tire from sidewall to sidewall. Observed on the vehicle placard or owner's manual and recorded by the interviewer.

Attributes: xx.xx   section width in inches

SAS Name: V12_HR   **SAS Label:** manuf. rec. flotation tire rim diameter

Type: Numeric

Remarks: High Flotation tire format: rim diameter in inches, vehicle manufacturer's recommendation. Observed on the vehicle placard or owner's manual and recorded by the interviewer.

Attributes: xx   diameter in inches

SAS Name: V12_NW   **SAS Label:** manuf. rec. numeric tire width

Type: Numeric

Remarks: Numeric tire format: section width component, in inches, vehicle manufacturer's recommendation. The outer width of an inflated new tire from sidewall to sidewall. Observed on the vehicle placard or owner's manual and recorded by the interviewer.

Attributes: xx.xx   section width in inches

Vehicle Level Variable Attributes

SAS Name: V12_NR   SAS Label: manuf. rec. numeric tire rim diameter

Type: Numeric

Remarks: Numeric tire format: rim diameter in inches, vehicle manufacturer's recommendation. Observed on the vehicle placard or owner's manual and recorded by the interviewer.

Attributes: xx   diameter in inches

SAS Name: V12_OTH   SAS Label: manuf. rec. other tire size

Type: Character

Length: 20

Remarks: Other tire format, tire size, vehicle manufacturer's recommendation. When tire sizes could not be classified into one of the three types discussed above they were grouped into the "other" category. Observed on the vehicle placard or owner's manual and recorded by the interviewer.

Attributes: xxxxxxxxxxxxxxxxxxxx   other tire size

SAS Name: V13   SAS Label: manuf. rec. pressure, front, cold

Type: Numeric

Remarks: Manufacturer recommended tire pressure, front, cold. Observed on the vehicle placard or owner's manual and recorded by the interviewer. If the placard/manual did not specify hot or cold and showed one pressure it was assumed to be cold.

Attributes: xx   psi

SAS Name: V14   SAS Label: manuf. rec. pressure, rear, cold

Type: Numeric

Remarks: Manufacturer recommended tire pressure, rear, cold. Observed on the vehicle placard or owner's manual and recorded by the interviewer. If the placard/manual did not specify hot or cold and showed one pressure it was assumed to be cold.

Attributes: xx   psi

Vehicle Level Variable Attributes

SAS Name: **V15**   **SAS Label: manuf. rec. pressure, front, hot**

Type: Numeric

Remarks: Manufacturer recommended tire pressure, front, hot. Observed on the vehicle placard or owner's manual and recorded by the interviewer.

Attributes: xx   psi

SAS Name: **V16**   **SAS Label: manuf. rec. pressure, rear, hot**

Type: Numeric

Remarks: Manufacturer recommended tire pressure, rear, hot. Observed on the vehicle placard or owner's manual and recorded by the interviewer.

Attributes: xx   psi

Tire Level Variable Attributes

SAS Name: CASENO   **SAS Label:** Case number

Type: Character

Length: 8

Remarks: This variable is a unique identifier for each vehicle in the data file. It is an 8-character field. The first two characters represent the PSU for the case, the second pair of characters represents the site in the PSU, the third pair represents the day the interview was conducted, and the last pair represents the individual vehicle for the PSU/Site/Day combination. The variable is on both the Vehicle and Tire level data sets and is used to link the two sets.

SAS Name: POSITION   **SAS Label:** observed tire position on vehicle

Type: Character

Length: 2

Remarks: Position of the tire on the vehicle

Attributes:
- LF   Left Front
- LR   Left Rear
- RR   Right Rear
- RF   Right Front

SAS Name: T_MANF   **SAS Label:** observed tire manufacturer

Type: Character

Length: 20

Remarks: Tire Manufacturer. Observed and recorded by interviewer from tire sidewall.

Attributes: xxxxxxxxxxxxxxxxxxxx Manufacturer name

NASS Tire Pressure Special Study

Tire Level Variable Attributes

SAS Name: T_TYPE **SAS Label:** observed tire size format

Type: Numeric

Remarks: Tire size format, observed tire. There are three commonly found tire size classifications for the types of vehicles in this study, each with its own format for presenting various size components. Tire size types that could not be easily categorized were placed in the "other" category. See discussion of tire sizes in the Introduction for more information.

Attributes:
1. Metric format (example "P205/75/R/14")
2. Light Truck High Flotation format (example "31X10.50 R15")
3. Light Truck Numeric format (example "8.75/R/16.5")
4. Other

SAS Name: T_C **SAS Label:** observed tire construction

Type: Character

Length: 1

Remarks: Construction type, all tire formats, observed tire. Observed on the tire sidewall and recorded by the interviewer.

Attributes: R Radial ply tire

SAS Name: T_SR **SAS Label:** observed tire speed rating

Type: Character

Length: 1

Remarks: Metric tire format: speed rating component, observed tire. The rating indicates the top speed for which the tire is certified. Observed on the vehicle sidewall and recorded by the interviewer.

Attributes:
- H   130 mph
- S   112 mph
- T   118 mph
- V   150 mph
- Z   over 150 mph

Tire Level Variable Attributes

SAS Name: **T_MT**   **SAS Label: observed metric tire type**

Type: Character

Length: 2

Remarks: Metric tire format: vehicle type component, observed tire. This indicator is omitted from many metric tire formats. Observed on the tire sidewall and recorded by the interviewer.

Attributes:
- P   Passenger Car Tire
- LT   Light Truck Tire
- T   Temporary Tire (spare)

SAS Name: **T_MW**   **SAS Label: observed metric tire width**

Type: Numeric

Remarks: Metric tire format: section width component, in millimeters, observed tire. The outer width of an inflated new tire from sidewall to sidewall. Observed on the tire sidewall and recorded by the interviewer.

Attributes: xxx   section width in millimeters

SAS Name: **T_MA**   **SAS Label: observed metric tire aspect ratio**

Type: Numeric

Remarks: Metric tire format: aspect ratio component, observed tire. The ratio between tire height and width. For example, an aspect ratio of 75 indicates a tire section height 75% as great as the width. Observed on the tire sidewall and recorded by the interviewer.

Attributes: xx   aspect ratio indicator

SAS Name: **T_MR**   **SAS Label: observed metric tire rim diameter**

Type: Numeric

Remarks: Metric tire format: rim diameter in inches, observed tire. Observed on the tire sidewall and recorded by the interviewer.

Attributes: xx   diameter in inches

Tire Level Variable Attributes

SAS Name: **T_HD**   **SAS Label: observed flotation tire diameter**

Type: Numeric

Remarks: High Flotation tire format, observed tire. Observed on the tire sidewall and recorded by the interviewer.

Attributes: xx   diameter in inches

SAS Name: **T_HW**   **SAS Label: observed flotation tire width**

Type: Numeric

Remarks: High Flotation tire format: section width component, in inches, observed tire. The outer width of an inflated new tire from sidewall to sidewall. Observed on the tire sidewall and recorded by the interviewer.

Attributes: xx.xx   section width in inches

SAS Name: **T_HR**   **SAS Label: observed flotation tire rim diameter**

Type: Numeric

Remarks: High Flotation tire format: rim diameter in inches, observed tire. Observed on the tire sidewall and recorded by the interviewer.

Attributes: xx   diameter in inches

SAS Name: **T_NW**   **SAS Label: observed numeric tire width**

Type: Numeric

Remarks: Numeric tire format: section width component, in inches, observed tire. The outer width of an inflated new tire from sidewall to sidewall. Observed on the tire sidewall and recorded by the interviewer.

Attributes: xx.xx   section width in inches

Tire Level Variable Attributes

SAS Name: T_NR **SAS Label: observed numeric tire rim diameter**

Type: Numeric

Remarks: Numeric tire format: rim diameter in inches, observed tire. Observed on the tire sidewall and recorded by the interviewer.

Attributes: xx   diameter in inches

SAS Name: T_OTH **SAS Label: observed other tire size**

Type: Character

Length: 20

Remarks: Other tire format, tire size, observed tire. When tire sizes could not be classified into one of the three types discussed above they were grouped into the "other" category. Observed on the tire sidewall and recorded by the interviewer.

Attributes: xxxxxxxxxxxxxxxxxxxx other tire size

SAS Name: MAXP **SAS Label: observed tire maximum pressure**

Type: Numeric

Remarks: Maximum cold pressure, observed tire. Observed on the tire sidewall and recorded by the interviewer.

Attributes: xx   psi

SAS Name: MSP **SAS Label: observed tire measured pressure**

Type: Numeric

Remarks: Measured tire pressure, observed tire. Measured and recorded by the interviewer. The maximum pressure that could be measured by the gauges used in the fieldwork was 60 psi, therefore the value of 60 represents 60 psi or higher.

Attributes: xx   psi

Tire Level Variable Attributes

SAS Name: **TEMP**   **SAS Label: observed tire temperature**

Type: Numeric

Remarks: Measured tire temperature, observed tire. Measured and recorded by the interviewer. Temperature was measured at the juncture of the tire tread and the sidewall, in line with the valve stem.

Attributes: xxx   degrees Fahrenheit

SAS Name: **TREAD**   **SAS Label: observed tire tread depth**

Type: Numeric

Remarks: Tread depth, in 32nds of an inch, observed tire. Measured and recorded by the interviewer.

Attributes: xx   32nds of an inch

www.ingramcontent.com/pod-product-compliance
Lightning Source LLC
Chambersburg PA
CBHW081824170526
45167CB00008B/3540